DYLAN TAUBER

PHOTOGRAPHER/COMPUTER ARTIST

To order additional copies of this book, contact:
Xlibris
1-888-795-4274
www.Xlibris.com
Orders@Xlibris.com

Print information available on the last page.

Rev. date: 08/16/2019

DoubleMirrors

Mankind's inexhaustible quest for truth/god/love or however else one wishes to refer to it, can be accounted for by what I call my Double Mirror theory. Like animals, our senses perceive a physical world; light reflects off tangible objects, and into the retina, sending a chemical message that is interpreted by the brain. This reflection causes other chemical functions leading to all of the natural behavior of animals. The chimpanzee sees the peanut and eats iti. What makes us different, however, is that we reflect on our existence, analyzing our sense perception from a distance. This ability to remember, to reflect on the past, in effect, provides a second "mirror" which combines with the first reflection to produce the illusion of infinity. We are like a chimpanzee living between two parallel mirrors, constantly viewing not only the peanut in front of him, but the reflection of the peanut and the reflection of the reflection and so on. As a result, we humans do not merely eat the peanut, but ponder why we are eating the peanut, what the world would be like if it were not for the peanut, and how we should properly treat the peanut, etc.. Like the illusion of infinity is produced by parallel mirrors, we construe notions of forever, immortality, the soul, god, and spirituality, etc. Once perceived, the illusion of infinity takes on a life of its own, and becomes a force infinitely more powerful than the mortal through which it was perceived. A computer can produce advanced technology but only man can perceive God.

The two mirrors produce a duality in our minds. This duality becomes metaphysical and therefore irrational, and as a result, it becomes difficult if not impossible to translate this experience into the limiting framework of rational words. It is impossible to rationally analyze double mirrors while allowing the idea to retain its infinite quality.

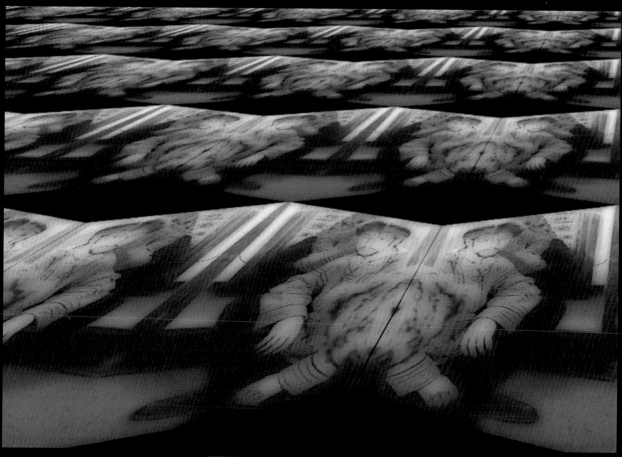

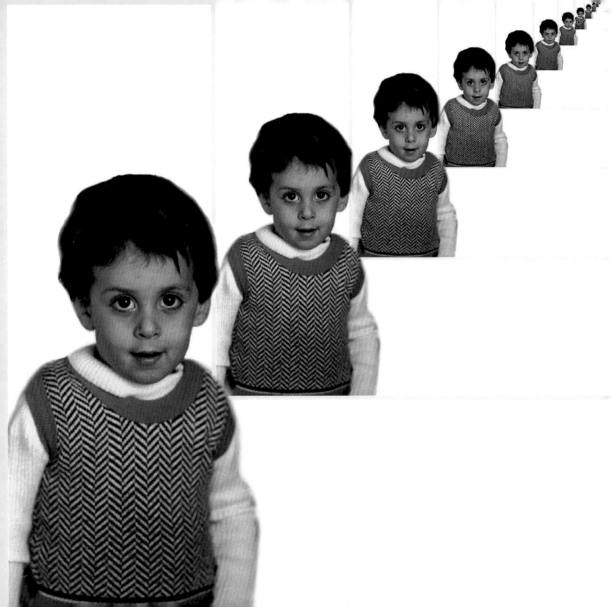

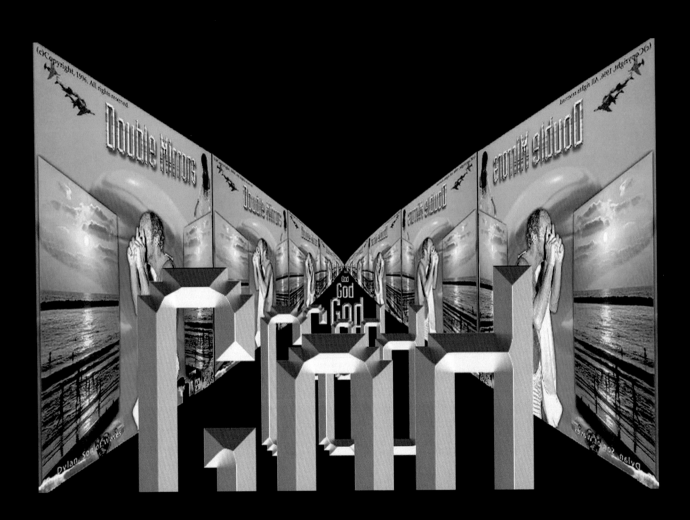

Dolphins

Dolphins

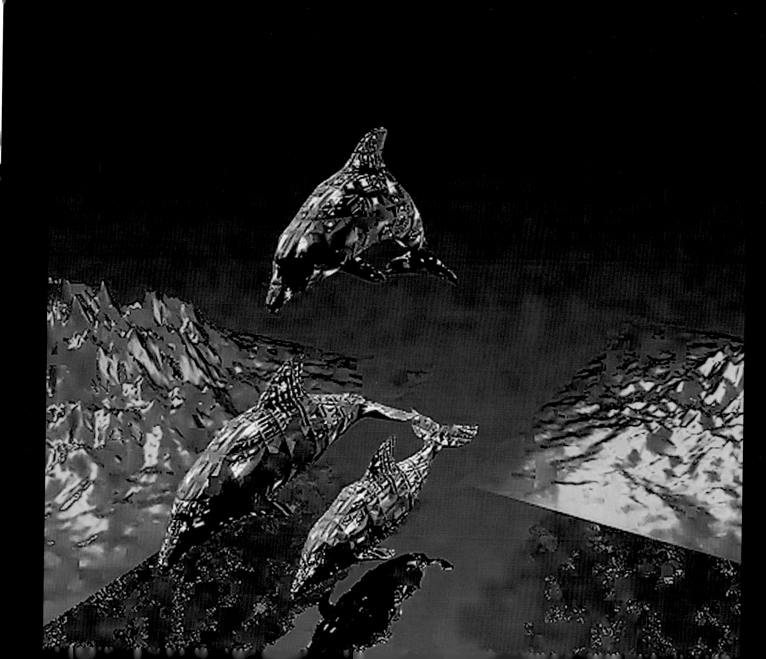

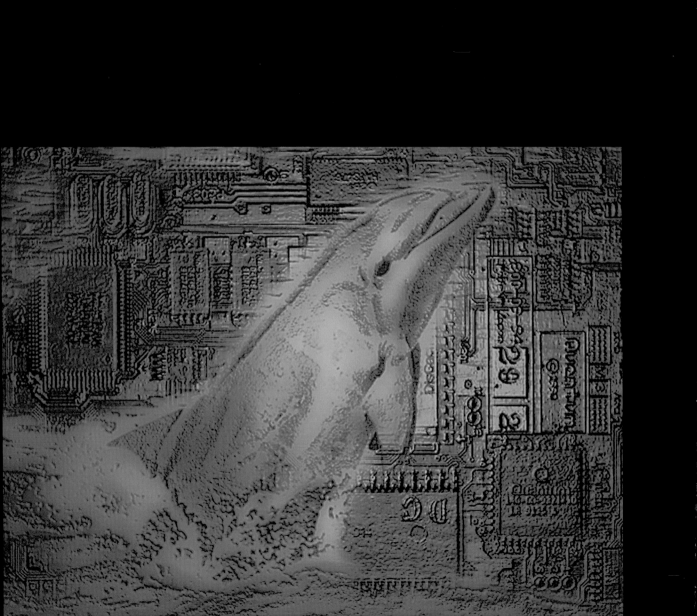

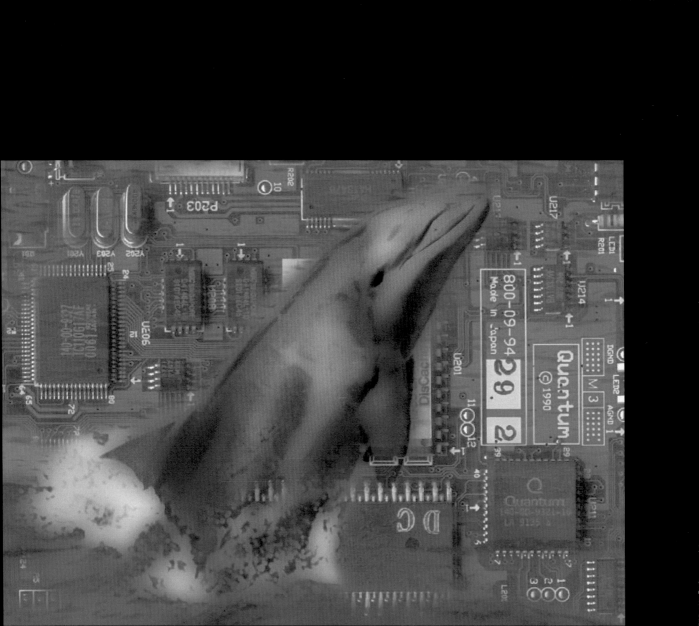

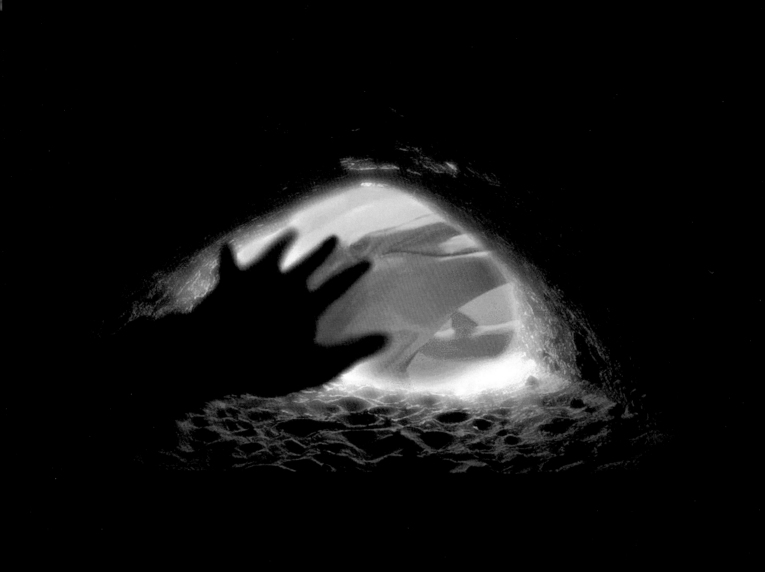

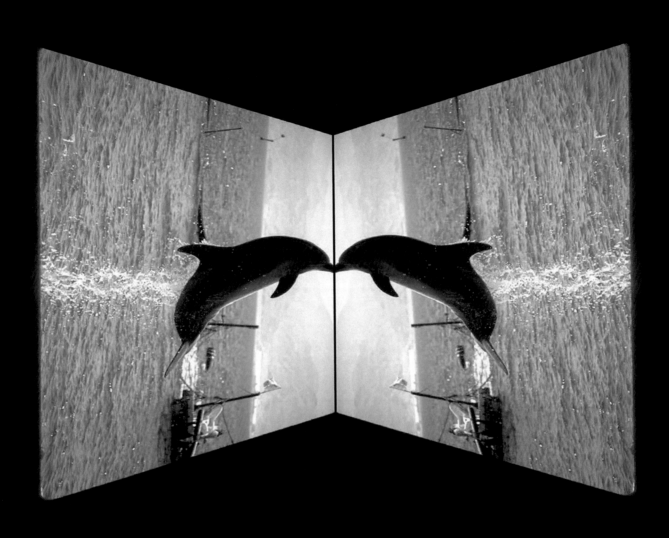

Dylan Tauber

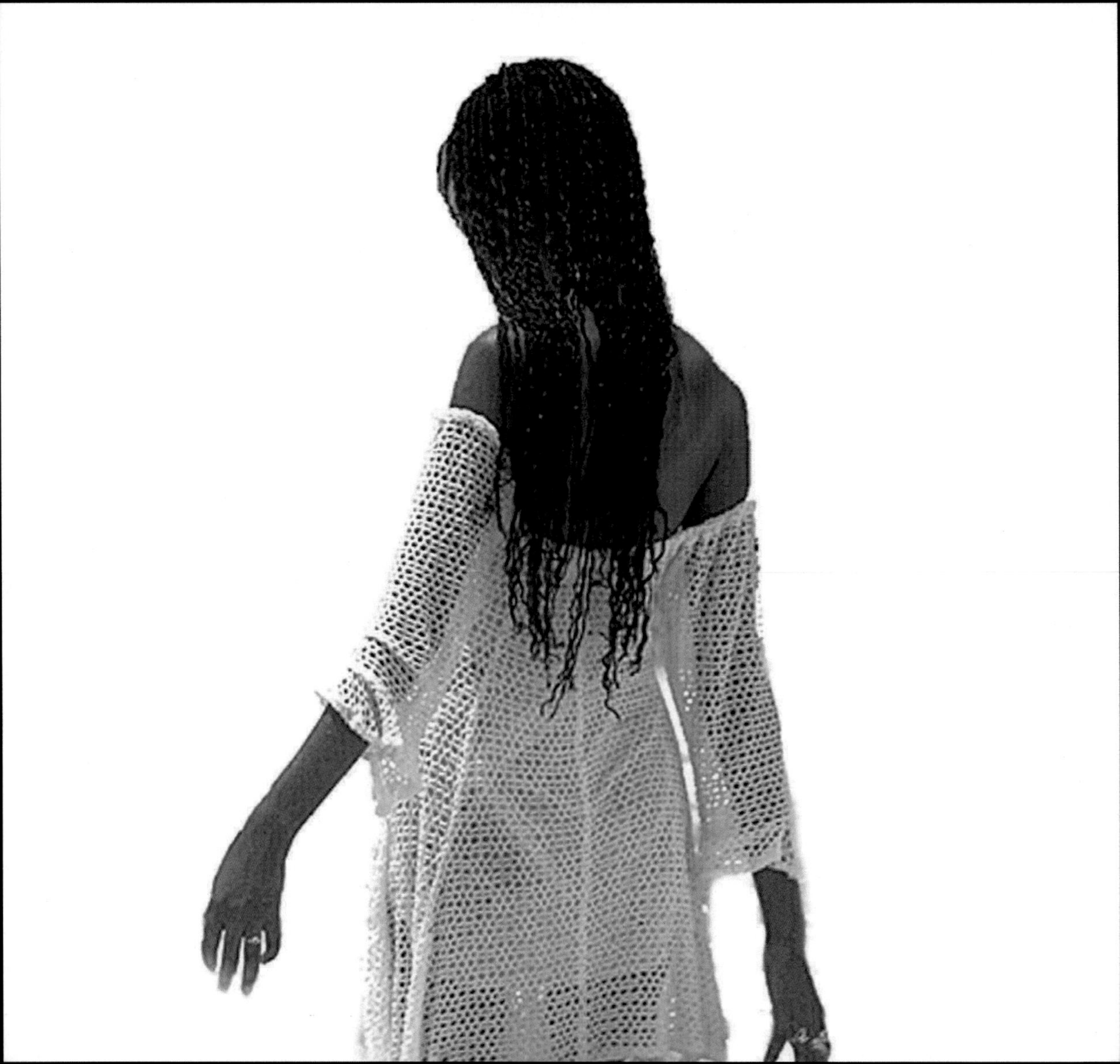

Ethiopian Woman

NYC

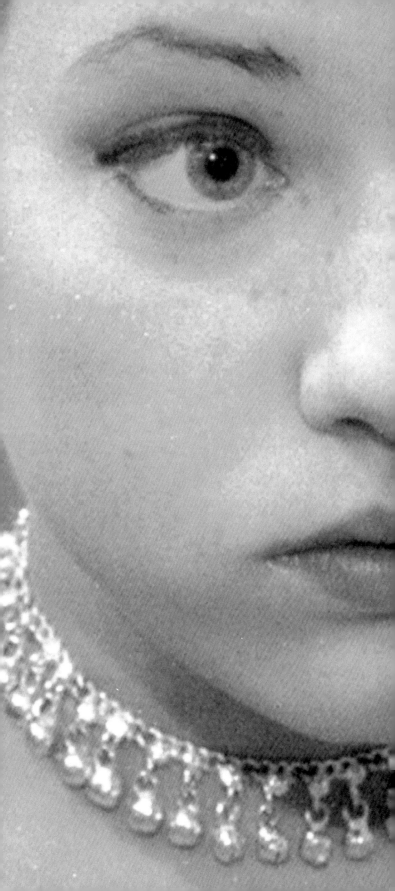

Printed in the United States
By Bookmasters